S0-DUU-088

NO LONGER PROPERTY OF
GLENDALE LIBRARY,
ARTS & CULTURE DEPT.

Ecosystems Research Journal

Rocky Mountains Research Journal

Natalie Hyde

Author: Natalie Hyde

Editors: Sonya Newland, Kathy Middleton

Design: Rocket Design (East Anglia) Ltd

Cover design: Margaret Amy Salter

Proofreader: Angela Kaelberer

Production coordinator and prepress technician: Margaret Amy Salter

Print coordinator: Margaret Amy Salter

Consultant:

Written and produced for Crabtree Publishing Company by White-Thomson Publishing

Front Cover:

Title Page:

Photo Credits:

Cover: All images from Shutterstock

Interior: Ron Dixon: p. 4; iStock: p. 8t Maxfocus, p. 9bl milehightraveler, p. 17t serg3d, p. 23b SrdjanPav, p. 24b benoitb, p. 25t milehightraveler, p. 26t Donyanedomam, p. 26b pukrufus; Shutterstock: pp. 4–5 Bjoern Alberts, p. 5 Protasov AN, p. 6 Christopher Jackson, p. 7t Scenic Shutterbug, p. 7b ajdebre, p. 8b Tom Grundy, p. 9t Tom Grundy, p. 9br Martina_L, p. 10 WitGorski, p. 11t Chase Dekker, p. 11b Inger Eriksen, p. 12t Sharon Day, p. 12b photogal, p. 13b Bildagentur Zoonar GmbH, p 13t tab62, p. 14t Rob Crandall, p. 14b Hein Nouwens, p. 15 silky, p. 16t Mark Goble, p. 17b Don Mammoser, p. 18t Lissandra Melo, p. 18b Ronnie Chuam, p. 19t Jeff Whyte, p. 19b Galina Savina, p. 20 Lissandra Melo, p. 21 Tom Reichner, p. 22t Ryszard Stelmachowicz, p. 22m Zovteva, p. 22b jorik, p. 23t Paolo Bona, p. 24t BGSmith, p. 25b Benny Marty, p. 27t Tom Reichner, p. 27b Pi-Lens, p. 28t Tomas Kulaja, p. 28b Constantine Androsoff, p. 29; Wikimedia: p. 16b.

Library and Archives Canada Cataloguing in Publication

CIP available at the Library and Archives Canada

Library of Congress Cataloging-in-Publication Data

Names: Hyde, Natalie, 1963- author.
Title: Rocky Mountains research journal / Natalie Hyde.
Description: New York, New York : Crabtree Publishing Company, 2018. |
Series: Ecosystems research journal | Includes index.
Identifiers: LCCN 2017029305 (print) | LCCN 2017030426 (ebook) ISBN 9781427119315 (Electronic HTML) |
ISBN 9780778734710 (reinforced library binding : alkaline paper)
ISBN 9780778734963 (paperback : alkaline paper)
Subjects: LCSH: Rocky Mountains--Environmental conditions--Research--Juvenile literature. | Biotic communities--Research--Rocky Mountains--Juvenile literature. | Ecology--Research--Rocky Mountains--Juvenile literature. | Rocky Mountains--Description and travel--Juvenile literature.
Classification: LCC GE160.R58 (ebook) |
LCC GE160.R58 H93 2018 (print) | DDC 577.0978--dc23
LC record available at https://lccn.loc.gov/2017029305

Crabtree Publishing Company

www.crabtreebooks.com 1-800-387-7650

Printed in Canada/082017/EF20170629

 In Canada: We acknowledge the financial support of the Government of Canada through the Canada Book Fund for our publishing activities.

Published in Canada
Crabtree Publishing
616 Welland Ave.
St. Catharines, Ontario
L2M 5V6

Published in the United States
Crabtree Publishing
PMB 59051
350 Fifth Avenue, 59th Floor
New York, New York 10118

Published in the United Kingdom
Crabtree Publishing
Maritime House
Basin Road North, Hove
BN41 1WR

Published in Australia
Crabtree Publishing
3 Charles Street
Coburg North
VIC, 3058

Contents

Mission to the Rockies

I received an exciting email today! The Mountain Research Group wants to know how the **ecosystems** in the Rocky Mountains are being affected by Earth's warming temperatures. They have asked me to travel across the entire range. The mountains stretch along the western coast of Canada and the United States. My report will cover the mountain range's many different landscapes, resources, plants, and animals.

Mountain ranges have different ecosystems. The types of plants and animals change from the foothills to the peaks. Valleys are usually grasslands. On the slopes of the mountains are pine forests. At the top there are only a few hardy trees. Scientists face many challenges protecting each of the different areas.

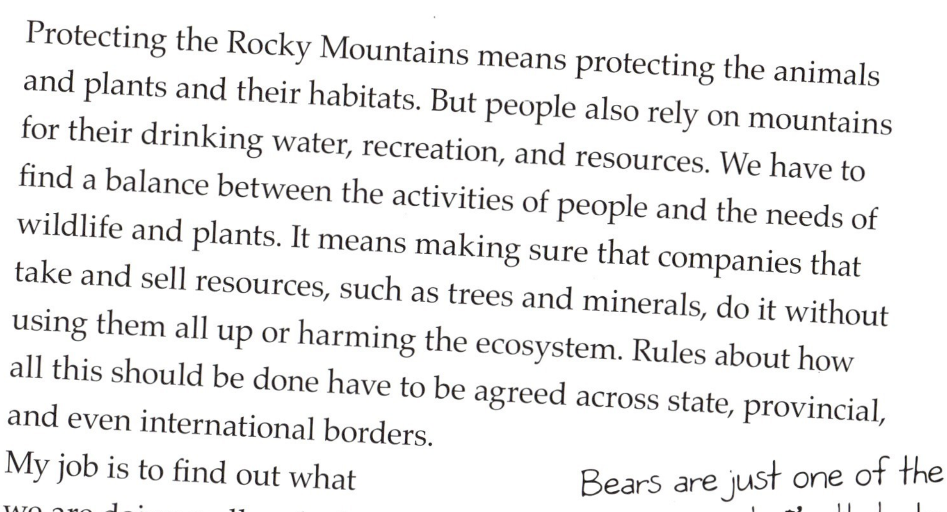

Protecting the Rocky Mountains means protecting the animals and plants and their habitats. But people also rely on mountains for their drinking water, recreation, and resources. We have to find a balance between the activities of people and the needs of wildlife and plants. It means making sure that companies that take and sell resources, such as trees and minerals, do it without using them all up or harming the ecosystem. Rules about how all this should be done have to be agreed across state, provincial, and even international borders.

My job is to find out what we are doing well and where we need to do more work.

Bears are just one of the many animals I'm likely to see on my trip through the Rockies.

Field Journal: Day 1

Boulder, Colorado to Rocky Mountain National Park

The first stop on my journey is Rocky Mountain National Park. This is one of many national parks in the Rocky Mountain range. The park system protects many different ecosystems. Each ecosystem supports a mixture of plant and animal life. I hiked through a section called Horseshoe Park, exploring the river valley and meadows. Then I climbed the mountain slopes, which were covered in trees such as lodgepole pines.

I kept my eyes open for bobcats. They make their home in the edges of the forests in this area.

In 1976, **UNESCO** gave Rocky Mountain National Park a special title. It was named one of the first World Biosphere Reserves. These reserves are protected areas. The park managers try to find a balance here. The goal is to protect but still use the park. Many activities are allowed. Visitors can hike, camp, and fish. These activities raise money to pay for programs but they also bring more people. This can cause damage to the fragile ecosystem. Special rules for visitors help protect the animals and plants that live there.

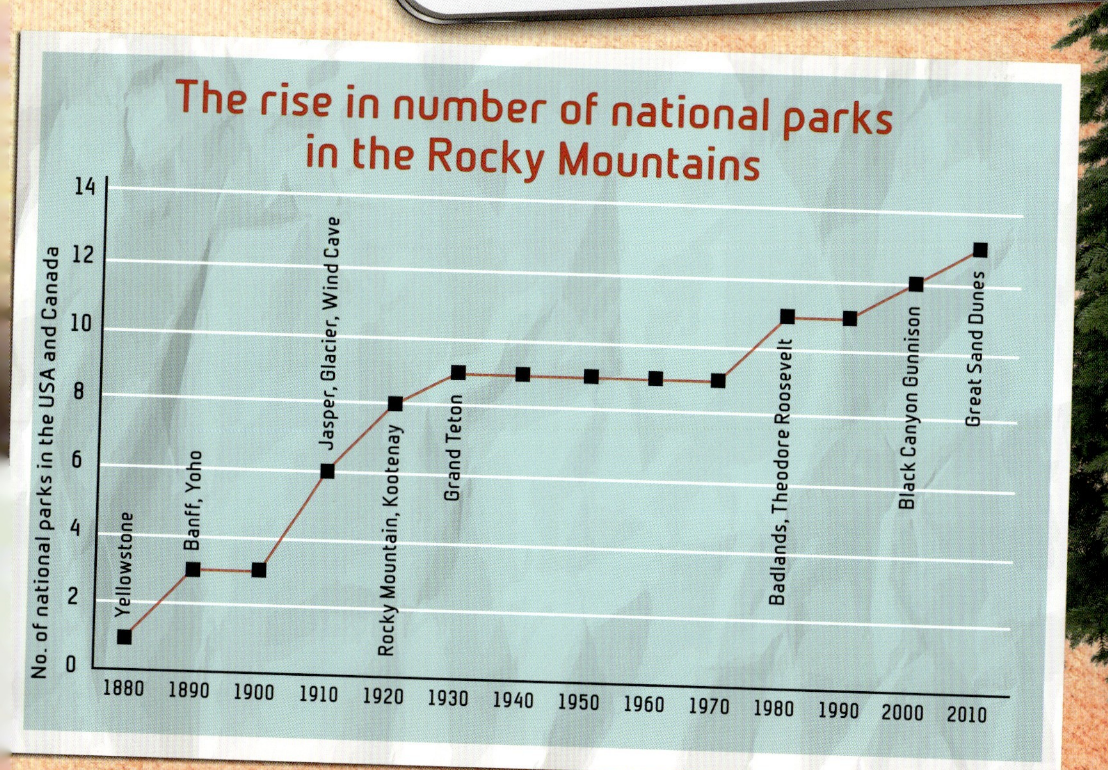

Field Journal: Day 2

Sinks Canyon State Park, Wyoming

I moved on to Sinks Canyon. This state park in Wyoming features a natural phenomenon. The Popo Agie River runs through the park. It disappears into a limestone hole called a **sink** and shows up again about 1,300 feet (400 meters) away in a pool called a **rise**. The journey above ground only takes a few minutes, but the water disappears underground for hours. Scientists used dye to figure out that it flowed through underground caves. Watching the river reminded me how important water is as a resource.

Middle Fork Falls is the source of the Popo Agie River. The water from the mountains cascades over large rocks, then drops into Sinks Canyon.

A snowpack, or large sheet of thick snow, sits high in the mountain peaks. This mass of snow stores water. The snowpack melts in the spring, and the water trickles down to join other streams. These drain into rivers that flow all the way to the ocean. It is vital to keep the water pure along the way because it is used by thousands of plants, animals, and humans. There are strict park rules on how to dispose of garbage, dishwater, and food scraps to avoid **polluting** the water. People use the snowpack's meltwater for drinking, farming, and industry. **Climate change** is making average temperatures higher. This could mean smaller snowpacks and less meltwater. This can lead to **droughts** in summer.

A snowpack high in the mountains.

natstat STATUS REPORT ST456/part B

Name: Boreal toad (*Bufo boreas boreas*)

Description:
These toads are brown or green, with a stripe down their backs. They do not make any mating calls like other toads. They eat grasshoppers, flies, beetles, and mosquitos. They like to live in higher elevations in **wetlands**. Many toads are suffering and dying from a skin disease called chytrid fungus.

Attach photograph here

Threats:
Chytrid fungus.

Status:
Threatened.

Field Journal: Day 3

Buildings in Jackson Hole are not allowed to be too tall. This makes it safer for planes. It also keeps the landscape more natural for wildlife.

Jackson Hole and Grand Teton National Park, Wyoming

I flew in to the town of Jackson Hole, which lies in a valley with the same name. The valley floor is made up of meadow and wetlands. The foothills are covered in rocky outcrops. Long ago, this was a prime spot for **trappers** hunting beaver. Now the town is a popular tourist destination because Grand Teton National Park is only 5 miles (8 kilometers) away. The airport in Jackson Hole is the busiest in Wyoming. Limits on the noise levels from the planes and other rules have been put in place to reduce the impact of human activity on the ecosystem.

Winter visitors can take a horse-drawn sleigh ride to see elk up close in the refuge.

The National Elk Refuge lies right on the edge of the town. I spoke with the manager of the refuge. He told me there were problems when Jackson Hole started to grow because it blocked the elk's **migration** route. The refuge was built to solve this problem. Instead of moving on, the 75,000 migrating elk are cared for here before they return to Yellowstone Valley.

natstat STATUS REPORT ST456/part B

Name: Bison (*Bison bison*)

Description:
Bison are the largest land animal in North America. They can move very quickly even though they are big and bulky. A bison's hump is made of strong neck muscles. A bison moves snow and ice to uncover plants to eat by shaking its head from side to side. Its only predators, other than humans, are wolves and grizzly bears.

Attach photograph here

Threats:
Habitat loss.

Status:
Near threatened.

Field Journal: Day 4

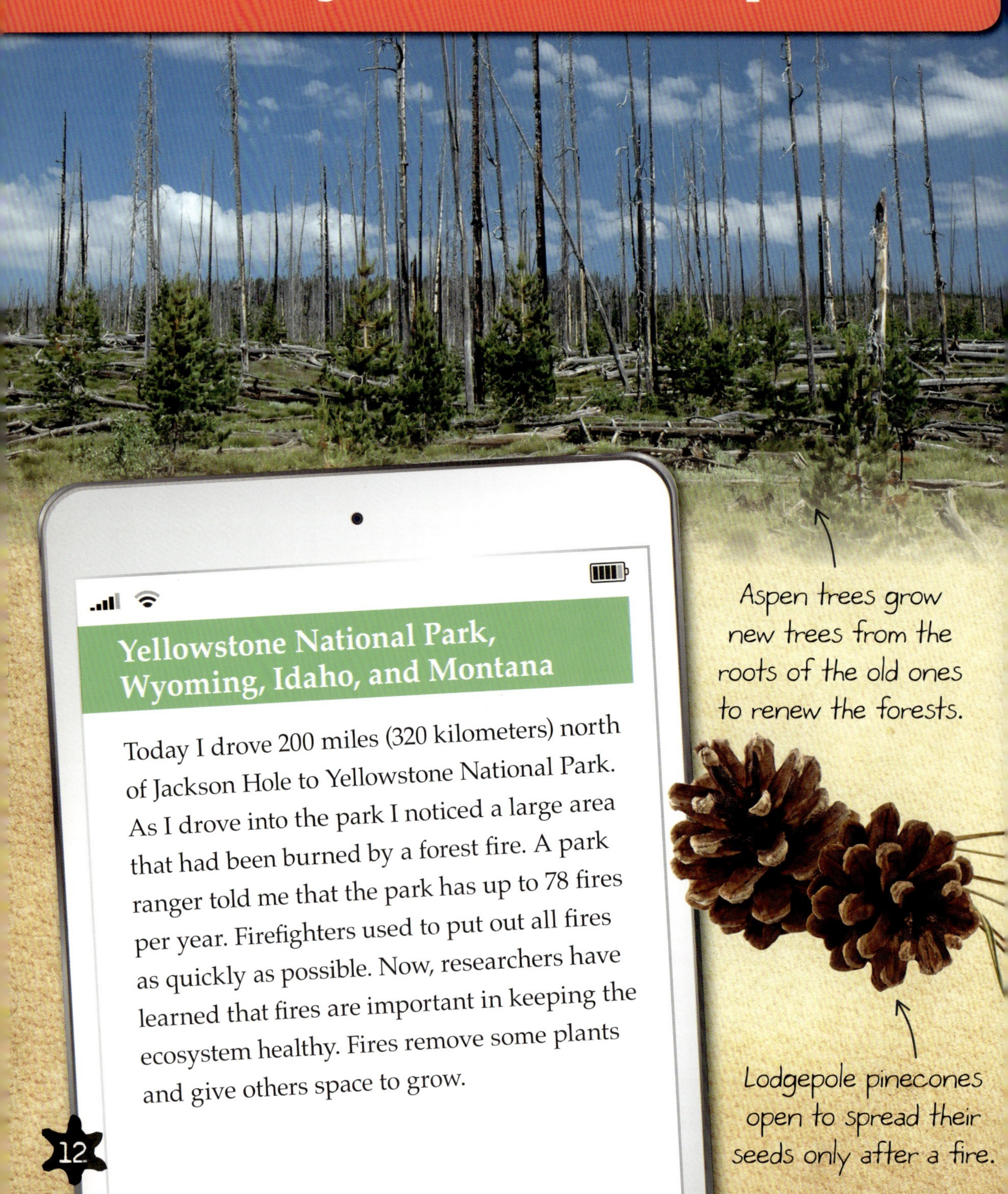

Yellowstone National Park, Wyoming, Idaho, and Montana

Today I drove 200 miles (320 kilometers) north of Jackson Hole to Yellowstone National Park. As I drove into the park I noticed a large area that had been burned by a forest fire. A park ranger told me that the park has up to 78 fires per year. Firefighters used to put out all fires as quickly as possible. Now, researchers have learned that fires are important in keeping the ecosystem healthy. Fires remove some plants and give others space to grow.

Aspen trees grow new trees from the roots of the old ones to renew the forests.

Lodgepole pinecones open to spread their seeds only after a fire.

Cutthroat trout.

Fires clear out deadwood, grasses, and shrubs. This allows other species to grow. Fires can also help control **invasive species**, which are a problem in Yellowstone. Houndstongue is a weed from Europe that is choking out **native** plants. It may have been brought into Yellowstone National Park in hay used for horses. The seeds attach to the coats of animals and are spread to other land areas as the animals move. Lake trout is an invasive animal. It is competing with the native cutthroat trout for food and resources. The cutthroat trout is a coldwater fish and is threatened by warming water temperatures. Park staff work to remove invasive animals or keep them in a small area.

Dead trees and dry grass are "fuel" for bigger forest fires. Park officials sometimes start fires themselves, known as "controlled burns." These fires can be kept under control and help lower the risk of wildfires.

The growth in size and impact of wildfires in North America

1985
1990
1995
2000
2005
2010
2015
USA
Canada
0 4 12 16
0 4 12 16
Area burned (in million acres)

It is easy to identify the invasive species houndstongue by its dark red flowers.

Field Journal: Day 5

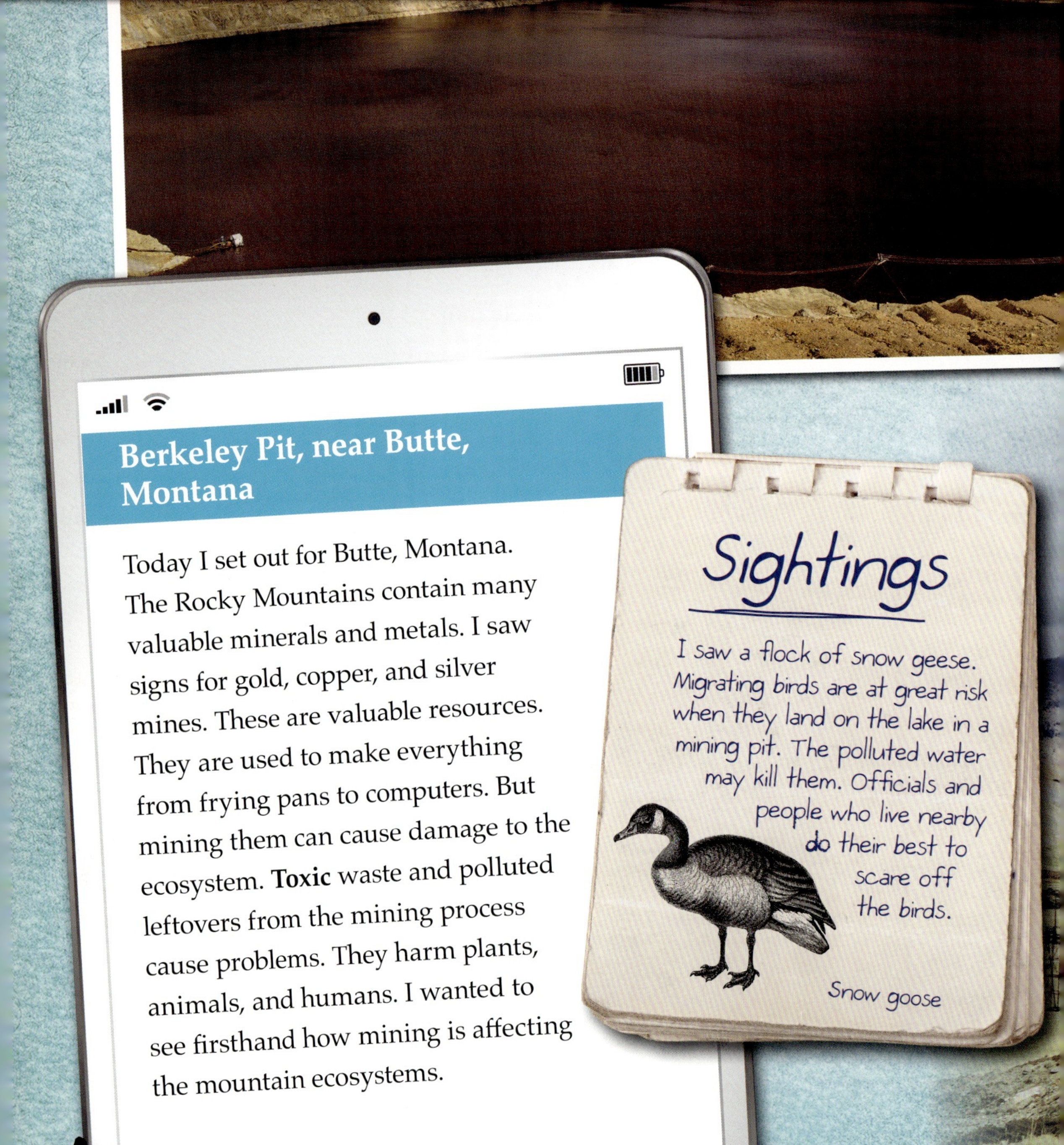

Berkeley Pit, near Butte, Montana

Today I set out for Butte, Montana. The Rocky Mountains contain many valuable minerals and metals. I saw signs for gold, copper, and silver mines. These are valuable resources. They are used to make everything from frying pans to computers. But mining them can cause damage to the ecosystem. **Toxic** waste and polluted leftovers from the mining process cause problems. They harm plants, animals, and humans. I wanted to see firsthand how mining is affecting the mountain ecosystems.

Sightings

I saw a flock of snow geese. Migrating birds are at great risk when they land on the lake in a mining pit. The polluted water may kill them. Officials and people who live nearby do their best to scare off the birds.

Snow goose

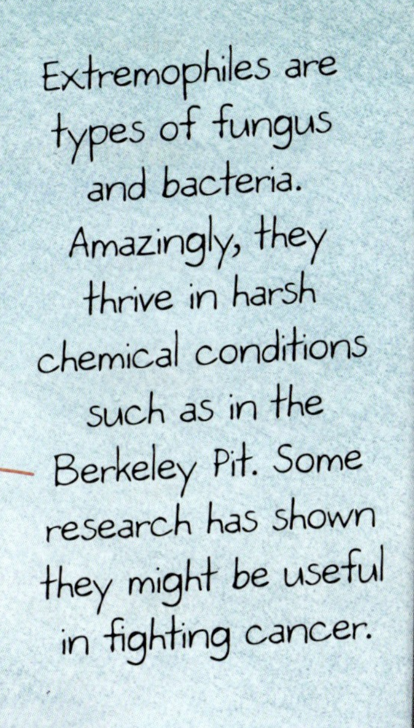

Extremophiles are types of fungus and bacteria. Amazingly, they thrive in harsh chemical conditions such as in the Berkeley Pit. Some research has shown they might be useful in fighting cancer.

I traveled to the site of a former copper mine. Berkeley Pit is an open pit about 1 mile long by half a mile wide (1.6 by 0.8 kilometers). It is abandoned now. Water in the pit is full of poisonous metals. There is so much of these metals, they can be mined directly from the water! The pit's water level is slowly rising. If it reaches the natural **water table**, it will flow into the **groundwater**. This will make the water dangerously polluted for animals and humans. The government is funding a cleanup of the site to stop this from happening.

Field Journal: Day 6

Waterton-Glacier International Peace Park.

Sightings

I was excited to see some Waterton moonwort. This is the rarest plant in Waterton National Park, and cannot be found anywhere else in the world.

Waterton moonwort

Waterton-Glacier International Peace Park, Montana/Alberta

The Rocky Mountains do not stop at borders. Neither do their unique and important ecosystems. Migrating animals choose ranges carefully. They look for a supply of food, water, and shelter. They do not care about borders. So how do we protect creatures and habitats that cross international lines? Canada and the United States built the Waterton-Glacier International Peace Park with a joint mission. The two countries share responsibility for this area.

Logging is illegal inside the park. But things are different outside the park boundaries. Forests are being stripped of trees for lumber, and the land is being used for other purposes. **Deforestation** causes habitat loss for many animals. To help solve this problem, the Canadian government created the Rocky Mountains Forest Reserve. This huge piece of land allows some industry and logging, but these activities are strictly limited. The rules protect the sources of the rivers from pollution. These headwaters supply most of the Canadian **prairies** with fresh water for agriculture and drinking.

natstat STATUS REPORT

ST456/part B

Name: Trumpeter swan (*Cygnus buccinator*)

Description:
This bird is the largest species of waterfowl. It is also one of the heaviest birds capable of flight. Adult males are completely white with a black bill. The trumpeter swan's cousin, the mute swan, has an orange bill. The call of a trumpeter swan sounds a bit like a trumpet, which is how it got its name. Mute swans live near humans, but trumpeter swans prefer remote wetlands with few people. They eat mostly water plants, grasses, and grains.

Attach photograph here

Threats:
Habitat loss, eating lead pellets used in guns for hunting, climate change.

Status:
Once near-threatened, now with conservation efforts, they are listed as least concern.

Field Journal: Day 7

Camping at Lake Louise, Banff National Park, Alberta

The needs of people and animals are very different. In the Rocky Mountains, people want space to ski, hike, swim, and fish. Animals need space to live, eat, and raise their young. It can be difficult to balance the two. I drove north to Banff National Park for a short camping trip at Lake Louise. I wanted to find out how the authorities here deal with this problem. I was surprised to see that the campsite was surrounded by an electric fence. The park ranger told me it was there for campers' protection because the animals are so wild—just how they should be!

In campsites here, campers are kept inside a fence and the bears wander free.

Even the highway into and out of the park has underpasses and overpasses so animals can cross safely. These passes are part of special routes through the valley called Wildlife Corridors.

Four Canadian national parks in the Rocky Mountains promote the "Bare Campsite" program. The idea is never to leave anything out in a campsite. In that way they do not attract wild animals, especially bears. All food, cooking utensils, garbage, dishes, and drinks must be kept locked up when people are away from the campsite. Park rangers do not want the bears to become less afraid of humans. Bears that get used to eating human food become a problem. They usually have to be moved to remote areas or be put down if they pose a danger to humans.

natstat STATUS REPORT ST456/part B

Name: Wolverine (*Gulo gulo*)

Description:
Wolverines are the largest members of the weasel family. They are very strong and can take down prey much larger than themselves. They are also scavengers. This means they eat the remains of other animals' kills. Wolverines live in forests and tundra, and need a large area to roam. Females build a den in the snow in February and usually have one to three "kits" in the spring.

Threats:
Loss of snowy habitat because of climate change, hunted for their pelts.

Status:
Vulnerable in 1996, now least concern as populations rebound.

Attach photograph here

Field Journal: Day 8

Tours of the glacier can educate people about how climate change is making the ice melt faster. However, more people and bus exhaust can make the problem worse.

Athabasca Glacier, Icefields Parkway, Alberta

I was excited to see my first glacier up close. The Athabasca Glacier in Alberta is a **tongue** of the large glacier system called the Columbia Icefields. These glaciers feed the streams and rivers that provide water for farms and towns. A sign pointed to where the edge of the glacier had been in 1890. It was a long walk from there to where it is now. I finally reached the glacier after walking a mile (1.6 kilometers). Scientists on site told me the glacier is melting rapidly. It is **receding** at least 16 feet (5 meters) every year. At this rate, scientists say it will have completely melted by the year 2100.

I left the glacier and drove up the Icefields Parkway. I stopped at the Bow Summit hiking trail. At the top I had a great view of Peyto Lake and Bow Lake. But even here I could see the effects of climate change. As temperatures rise, the **tree line** also moves farther up the mountain. This affects the highest ecosystem, called the alpine tundra. This ecosystem is in danger of disappearing. This is the habitat of bighorn sheep and pika. Pika are little mammals that do not thrive in high temperatures.

I almost stumbled into a crevasse. These deep open cracks in the glacier are created as the glacier moves and shifts. They can form in a matter of days and are extremely dangerous.

If temperatures continue to rise, the pika may be extinct in 100 years.

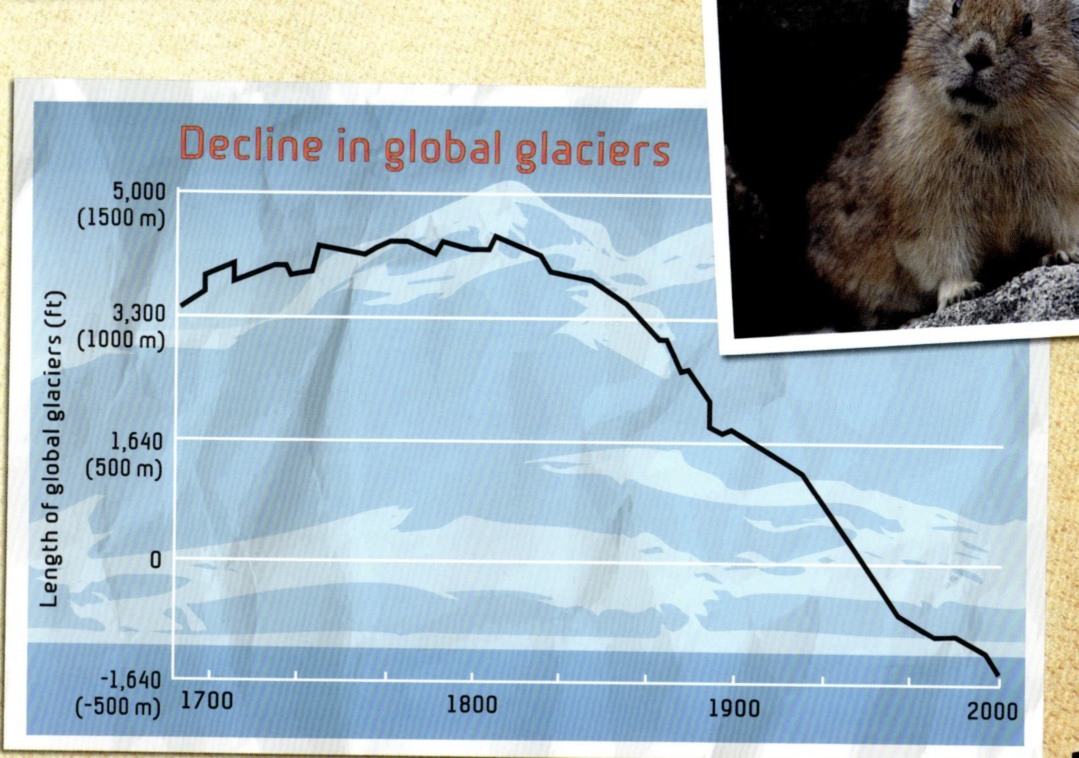

Field Journal: Day 9

Whistler Mountain, near Jasper, Alberta

Today I headed to the very top of Whistler Mountain. Luckily I did not have to hike all the way up! There is a tram that takes visitors 7,546 feet (2,300 meters) up the mountain. From there, I hiked about 30 minutes to the summit. There are no trees this high up. The ground is bare and rocky. I had hoped to see Mount Robson from the top. It is the highest point in the Canadian Rockies. But visibility was poor. I could only see about 37 miles (60 kilometers).

Many people think the air quality in the Rocky Mountains is very pure. But the quality can be lowered by pollution from cities and forest fire smoke. An increase in the air of a gas called **ozone** from car exhaust can make breathing difficult. My guide told me that power plants and fertilizers used for crops produce high levels of the gas nitrogen. Nitrogen is blown by the wind and deposited over a wide area by rain or snow. This can damage native plants and allow invasive plants to thrive.

Gases in car exhaust contribute to Earth's warming temperatures.

natstat STATUS REPORT ST456/part B

Name: Half-moon hairstreak butterfly (*Satyrium semiluna*)

Description:
This butterfly is usually small and drab. It has blackish-brown wings with black spots. It feeds on lupines and sagebrush. In a single year (2003–2004), numbers of half-moon hairstreak butterflies in Alberta parks fell from over 10,000 to under 250. Scientists think a late frost may have killed many breeding pairs.

Threats:
Habitat loss due to development and cattle ranching.

Status:
Endangered.

Attach photograph here

Field Journal: Day 10

Robson Valley.

Robson Valley, British Columbia

This beautiful and sunny day was perfect for a mountain bike ride. I went to meet with a research team in Robson Valley in British Columbia. As I biked I noticed the pine trees. Many of them had needles that were rust-colored rather than green. I found the team deeper in the forest. They were inspecting the trunks of lodgepole pines. I went to help, and I was told to look for pitch tubes in the tree trunks. These are tiny holes that female pine beetles bore into trees to lay eggs. The tree fights back by filling the tubes with **resin** to get rid of the pest. The rust color of the needles is a sign that the tree has been infested.

Sightings

I spotted a pileated woodpecker. This was a good sign because they prey on mountain pine beetles. As a group, these woodpeckers can eat up to 30 percent of a pine beetle colony.

Woodpecker

I was amazed how many trees had been affected. The researchers told me that the mountain pine beetle has already killed about half the lodgepole pines in British Columbia. These pests are now moving east. There have been outbreaks in Alberta. Milder winters seem to have helped the beetle survive. Disease is always a challenge in maintaining healthy ecosystems. The animals are suffering, too. Chronic wasting disease is a serious problem. It causes deer to lose weight until they die. It is spread by animals eating grass near infected dung. It was first found in Colorado, but it has spread to 23 states and two provinces.

Mountain pine beetle.

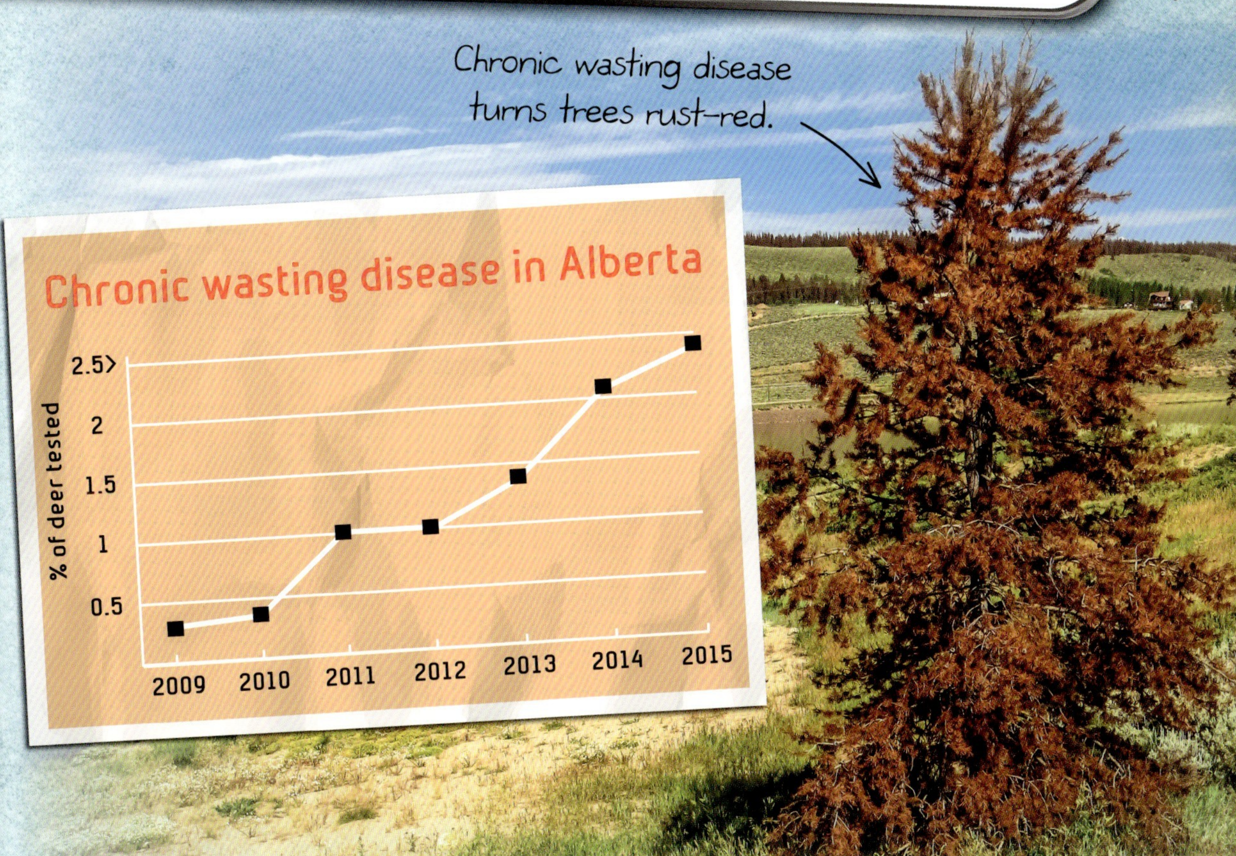

Chronic wasting disease turns trees rust-red.

Field Journal: Day 11

An avalanche path
down a mountainside.

Mountain climbing in Northern Rocky Mountains Provincial Park, British Columbia

I drove a long way before finally reaching the Northern Rocky Mountains. I was asked to join a team of researchers studying the effects of avalanches. I parked my car and found they had saddled up an extra horse for me. Horseback is the best way to go deep into the park because there are no roads. I got a good view of the land around me as I rode. Our team followed a trail along the base of some steep slopes. There we saw evidence of an avalanche. Soil and rocks pushed by a small avalanche had created a wall of debris. Meltwater had collected behind this wall and formed a small lake, called an impact pool.

Sightings

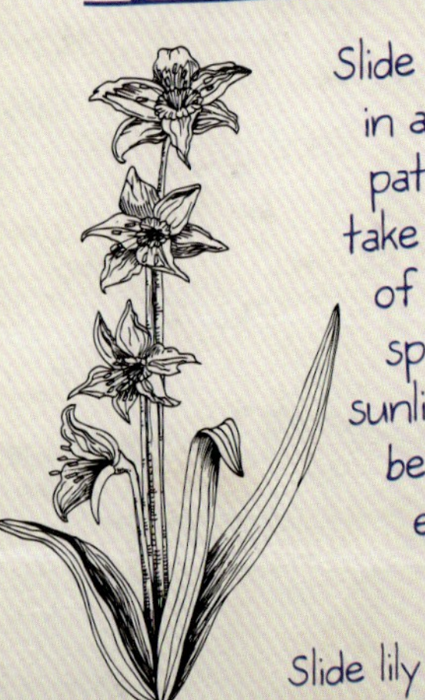

Slide lilies grow in avalanche paths. They take advantage of the open spaces and sunlight. Grizzly bears like to eat these plants.

Slide lily

Farther on we saw a path of destruction from an avalanche that happened in a previous winter. Trees were uprooted. Boulders lay all around. We stopped to investigate. The researchers showed me that even land cleared by avalanches serves a purpose for the animals in the habitat. Bears were digging here for roots and bulbs. I saw birds called ptarmigan using the paths made by the avalanche. These birds stay close to the ground to conserve their energy in the cold. Avalanches can be dangerous and destructive, but they are also a natural part of the changing mountain ecosystems.

Ptarmigan.

natstat STATUS REPORT ST456/part B

Name: *Stone sheep (Ovis dalli stonei)*

Description:
A subspecies of tinhorn sheep, stone sheep are brown with patches of white on their rumps. They have big, curved, yellow-brown horns. They eat many different plants in summer. They eat frozen grasses, mosses, and lichens in winter. They travel many miles to visit mineral licks in the spring. These deposits on the ground give sheep important minerals like salt and calcium. Their main predators are wolves, coyotes, and bears.

Attach photograph here

Threats:
Loss of habitat through mining, roads, and human activities.

Status:
Least concern.

Final Report

Report to: MOUNTAIN RESEARCH GROUP

OBSERVATIONS

My trip through the Rocky Mountains has been incredible. I learned that it is challenging to keep all the mountain range's many ecosystems healthy for both animals and humans.

FUTURE CONCERNS

Climate change plays a role in most of the changes happening to all of the Rocky Mountain ecosystems today. Warmer average temperatures are moving the tree line and the alpine habitat. This shrinks the habitat of cold-weather animals. Snowpacks and glaciers are melting at a quicker pace. This causes flooding in spring and drought in summer. The supply of fresh water is also threatened if the snowpacks and glaciers ever melt completely.

Nature renews itself through fires, floods, and avalanches.

Conservation Projects

Governments in the United States and Canada continue to expand the chain of national, state, and provincial parks. This helps protect more of the Rocky Mountains' fragile ecosystems. Governments also create laws to make sure that companies limit long-term damage to the environment when they take resources such as minerals and trees. There are rules that limit the amount of **greenhouse gases** industries can put into the air. However, these laws can change when governments change. That is why many countries are joining together to fight climate change. Countries that have signed the 2015 Paris Agreement on Climate Change have agreed to limit their greenhouse gases and spend more money on **clean energy** sources. If we take care of the many parts of this beautiful ecosystem now, we will be able to enjoy the Rocky Mountains for generations.

* What information can you get out of journal entries that you cannot get from a report?
* The graph on page 21 reflects what is happening to glaciers in the world, not just the Rocky Mountains. What can you learn from a graph, even if the information is not about the exact place or object or event that you are researching?
* Our scientist often includes "Sightings" of plants, animals, or landforms in her journal. Why and how are these useful to journal readers?

Learning More

BOOKS

Rocky Mountains by Scott A. Elias (Smithsonian Institution Scholarly Press, 2002)

The Rocky Mountains by Molly Aloian (Crabtree Publishing, 2011)

Unfolding Journeys: Rocky Mountain Explorer by Annie Davidson and Stewart Ross (Lonely Planet Global Limited, 2016)

WEBSITES

http://video.nationalgeographic.com/video/rm-amphibians
Learn about scientists researching climate change by looking at salamanders in the Rocky Mountains in this National Geographic video.

http://kids.britannica.com/comptons/article-207407/Rocky-Mountains-or-Rockies
Discover the four main sections of the Rocky Mountains, as well as their history, with Encyclopedia Britannica Kids.

http://video.nationalgeographic.com/video/news/151021-glacier-national-park-melting-vin
Look at evidence of shrinking glaciers with National Geographic.

Glossary & Index

clean energy energy from sources such as the wind, sun, or water, which will never run out

climate change a change in the normal weather conditions in an area over time, caused by pollution and other human activities

deforestation cutting down areas of forest to use the land for another purpose

drought a long period where there is less rain than usual in a particular region

ecosystem a community of plants and animals

greenhouse gas a gas such as carbon dioxide that contributes to global warming

groundwater water found underground in the spaces in soil, rock, and sand

invasive species plants or animals that move from their native area to a new area

migration movement from one region to another according to the seasons

native born in a particular area

ozone a colorless toxic gas

polluting making something like water or air dirty or unusable

prairie a large area of grassland

receding moving backward

resin a sticky substance discharged by some trees

rise a pool where underground water comes to the surface

sink a place where water drops below Earth's surface

tongue ice that flows away from the main trunk of a glacier

toxic poisonous or harmful

trappers people who catch animals in traps, often for their fur

tree line the point on a mountain or other high area above which no trees grow

UNESCO United Nations Educational, Scientific and Cultural Organization—an international organization that helps protect and preserve cultural sites around the world

water table the level below which the ground is saturated with water

wetlands areas of waterlogged land such as marshes or swamps